German U-Boat B
Yesterday and Today

The U-boat bunker in the south harbor of St. Nazaire.

Karl-Heinz and Michael Schmeelke

Schiffer Military/Aviation History
Atglen, PA

Bibliography and Acknowledgments

Neitzel, Sönke, Die deutschen Ubootbunker und Bunkerwerften, Bernard & Graefe Verlag, 1991.
Pallud, J. P., U-Boote-Les Bases, Heimdal, 1989.
Peron, Francois, Brest, sous l'occupation, Ouest France, 1981.
————, U-Boote im Einsatz, Podzun Verlag, 1970.
Pfefferle, Ernst, Kameraden zur See, Vol. 1-8, Altmannstein, n.d.
Farmbacher-Matthiae, Lorient, Prinz Eugen Verlag, 1956.
Herzog, Bodo, Deutsche U-Boote 1906-1966, Pawlag Verlag, 1990.
Zimmermann, R. H., Der Atlantikwall, Vol. 2, Schild Verlag, 1989.
KTB Royal Naval Air Service, Dunkerque 1917-18.
39/45 Magazine, Heimdal, Bayeux, 1984-94 issues.

Cover Picture

Cover artwork by Steve Ferguson, Colorado Springs, CO.

Acknowledgments

Our special thanks go to the following persons who, as eyewitnesses, gave us much encouragement and advice: Messrs. Karl Hoffmann (11th U-Flotilla), Rolf Leyse (16th MS Flotille), Karl Hartung (MAA 262), Karl Bach (MFAA 704), Alfred Uher (MFAA 807), as well as the French harbor workers who guided us through the bunkers in St. Nazaire, La Pallice and Bordeaux.

The lock bunker at St. Nazaire. After the harbor had become the target of a British commando operation on March 28, 1942, the bunkers were given numerous short-range defenses such as machine-gun nests, 5 cm grenade launchers, and 47 mm Pak (t) guns.

Translated from the German by Ed Force

Copyright © 1999 by Schiffer Publishing, Ltd.

Printed in China.
ISBN: 0-7643-0786-X

This book was originally published under the title, *Waffen Arsenal-Deutsche U-Boot Bunker Gestern und Heute* by Podzun-Pallas Verlag.

We are interested in hearing from authors with book ideas on related topics.

Published by Schiffer Publishing Ltd.
4880 Lower Valley Road
Atglen, PA 19310
Phone: (610) 593-1777
FAX: (610) 593-2002
E-mail: Schifferbk@aol.com
Please visit our web site catalog at www.schifferbooks.com
or write for a free catalog.
This book may be purchased from the publisher.
Please include $3.95 postage.
Try your bookstore first.

UB 90 sets out from Bruges on a mission, under the protection of naval aircraft.

Introduction

At the beginning of the year 1915, the German Admiralty staff believed the time had come when they could begin the U-boat war against their opponents' merchant fleets, although only 22 boats were available and ready for service. Until then, U-boats had only been used in operations against purely military targets. On February 4, 1915, Germany declared the sea area around the British Isles to be a war zone. To be sure, the hopes for a blockade in the spring and summer were not fulfilled, what with the much too small numbers of U-boats. After the sinking of the liner *Lusitania*, in which a number of American citizens lost their lives, the U-boat war against merchant ships was strongly limited again, on account of political pressure from the USA. Only after a German peace proposal had been rejected in 1917, did the unlimited U-boat war in the British war zone begin on February 1 of that year.

In the first five months alone, the German U-boats sank over 4.5 million tons of enemy shipping. When the USA entered the war in 1917, the British defensive actions could be improved greatly, and the airplane developed very quickly as a serious enemy of the U-boat. Still in all, the sinking statistics of the German Navy in 1917 remained at a monthly 600,000 tons. Only in 1918 did the tonnage sunk decrease to 370,000 tons per month. Of the total of 178 German U-boats of World War I, 178 were sunk by enemy action, 132 of them alone in the last two years of the war.

In order to prevent the German U-boats of the Flanders Flotilla from departing, the British sank several ships in the Bruges Canal in a commando operation in April 1918. To be sure, the blocking was only incomplete, and after just a few hours, the German U-boats could pass through the canal again.

On June 28, 1935, the German Navy's first new U-boat was launched at Kiel, and the numbers were to grow to over 1160 boats by the end of World War II. Because of the experience that had been gained during World War I, a bunkered dock was planned for every U-boat that returned from combat. The first U-boat bunker was built on the island of Helgoland in 1940; others arose in Hamburg and Kiel.

Enemy U-boats were also threatened with danger from the air. On July 6, 1918, a squadron of German naval airplanes surprised the two British submarines C 25 and E 51 off the British coast and damaged them both badly. While E 51 could still dive and escape, C 25 took several hits in her compressed-air tanks and was immediately incapable of diving. Despite her serious damage from machine-gun fire from the Hansa Brandenburg airplanes, they were not able to sink her.

The conquest of the Norwegian and French Atlantic coasts in 1940 brought a significant improvement to the German sea-war command's basis of operation, particularly for the U-boat war. From the harbors of the Atlantic coast and the Norwegian bases, the boats had a shorter and less dangerous approach course to the areas of operations. Naturally, this also allowed the numbers of U-boats in contact with the enemy to be increased.

Requirements for all sea operations were, of course, the suitable land support points, centered around the shipyards which always had to be ready for use by the ships sent to them.

The Commander of the U-boats (B.d.U.), Admiral Dönitz, had established the following standards. One third of the boats were to attack the enemy, one third were to be going or returning, and the last third were to be overhauled at the shipyards. As of June 1940, the Navy had the two harbors of Lorient and Brest expanded as U-boat bases. The two harbors already possessed large French naval arsenals, which could be taken over at once by the German Navy. St. Nazaire followed in 1941, and the Italian U-boats were assigned to Bordeaux as their supply base. With the increase of the U-boat war and the number of boats, the Navy set up further support points at La Pallice and Bordeaux. Until 1945, the naval shipyards of Brest, Lorient, St. Nazaire, La Pallice and Bordeaux carried out, in all, 1149 U-boat overhauls. In Norway, Bergen and Trondheim acquired larger shipyards with bunkers.

At the shipyards, though, not only did the normal maintenance work have to be carried out, but there were also additional Flak guns, radar-detecting apparatus and snor-

kels to be installed in the boats. Here the shipyards performed great improvisational work. In the autumn of 1940, the first enemy air raids on the U-boat support points began, and the civilian population had to suffer a great deal as well.

The RAF and the USAAF used specially developed bombs against the U-boat bunkers as of 1944. They were supposed to be capable of penetrating four meters of concrete and then explode in the bunker thanks to a delayed fuse. In practice, though, it was found that the bombs exploded after about two meters of concrete. The bombs tore holes up to eight meters in diameter in the decks of the U-boat bunker at Brest, but the explosive force was expended upward for the most part. The damage to the boats in the bunkers was done by falling concrete blocks, and only seldom by bomb splinters. With a deck thickness of about five meters, the interior of the bunker remained almost completely protected from pressure and splinter effects.

As the Allied air war became more and more severe, the German area began in 1943 to make plans to build U-boats in bomb-safe production facilities. In the same year, the construction of shipyard bunkers began in Bremen and Kiel. Others were planned for Hamburg and Wilhelmshaven. In November 1944, Dönitz promoted a large-scale building program that involved 178 new bunker spaces for U-boats in German harbors, but these never went beyond the planning stage. Despite massive bombing attacks, the Allies did not succeed in having a decisive effect on the German U-boats in their harbors. After the surrender, 159 commanders turned their boats over to the British, while 203 boats were sunk at sea by their own crews. More than 28,000 men of the German U-boat fleet fell during World War II.

The bunker in La Pallice during its construction in 1941. At that time the bunker included seven boxes and offered protection to nine U-boats.

Lorient in 1943: Large portions of the city were destroyed by Allied bombing raids, while the shipyards were scarcely damaged.

U 97 "Seepferd" (Seahorse) arrives at St. Nazaire in the summer of 1942. The boat was sunk in an air raid off Haifa on June 16, 1943; there were no survivors.

The Bruges U-Boat Base

Since December 1914, German U-boats of the Flanders Naval Corps operated from the harbors of Bruges and Ostende against warships in the sea zone around Britain and in the North Sea. In addition to British and French defensive measures such as mine barrages, U-boat nets, listening stations and searchlights, the greatest danger to the German U-boats came from the air.

The British Navy used steerable balloons, the so-called coastal airships, to watch the coastal sea areas. The balloons had radio sets, small searchlights and, for direct action, light bombs on board. The heavily armed Curtiss and Felixstowe flying boats patrolled over the open sea from 1917 on, and for the first time the spiderweb system was utilized. For it, the area was divided into geometrical sectors, and the flying boats searched systematically, sector by sector. On May 20 the Curtiss H 12 flying boats sank their first German U-boat, probably UB 70. The bombers from RNAS Dunkerque attacked the U-boats as they entered the harbors of Ostende and Bruges, and while they lay in the harbors.

The protection of the U-boats during their entry and exit was taken over by the German naval aircraft from the Ostende and Zeebrugge bases. To protect the boats in the harbors, the Navy built U-boat nests and, in August 1917,

UB 27 returns to the harbor or Bruges after laying mines along the British coast.

the building of the first U-boat bunker began in the northern torpedo-boat harbor at Bruges. This first nest built of reinforced concrete consisted of eight boxes, each 62 meters long and 8.80 meters wide. Despite many bombing raids on Bruges by the RNAS, the bunkers were never seriously damaged or even penetrated by a bomb.

In May 1918, a 230-pound bomb struck a shelter in Bruges harbor. The bomb exploded on impact on the 10-mm-thick sheet-steel roof of the facility. The U-boat under the roof was damaged in its superstructure and pressure system.

The U-boat bunker at Bruges, seen during construction at Bruges in the spring of 1918.

Right: A German high-sea U-boat of the Flanders Flotilla comes out of the bunker in Bruges.

Below: A VII C boat of the first U-Flotilla ties up to the Jean Bart Quay in Brest after being transferred from Kiel in July 1941.

Brest

On June 18, 1940 Brest had been occupied by the German Wehrmacht, and the rebuilding of the partially destroyed harbor was begun at once. A few weeks later, on August 22, the first German U-boat, U 65, already entered the harbor for a brief overhaul. In June 1941 the Navy began to establish the 1st U-Flotilla, named "Weddingen" after Kptl. Otto Weddingen, U-boat commander and Pour le Merite holder of World War I. In November the 9th U-Flotilla was established, also at Brest. At the beginning of 1941, the Organisation Todt began to build a U-boat bunker in the harbor of Brest. This structure, situated under the naval school, included ten drydocks and five wet boxes in its final form, and offered protection to a maximum of twenty boats. It measured 333 meters wide, 192 meters long and 17 meters high. The wet boxes, with a length of 115 and a width of 17 meters, could each hold three boats if the support point was overfilled, though in this case several meters of one U-boat's stern stuck out of the bunker. Drydocks 1 to 8, 96 to 99 meters long and 11 meters wide, were planned for one boat each, as were Docks 9 and 10, each 114 meters long. As of the summer of 1942, all the bunker spaces in Brest were ready for use. Although the two U-flotillas in Brest were equipped only with types VII C and VII D boats, boats of types IX and X sometimes came to be serviced, as did Japanese U-cruisers. An extension of the bunker by ten boxes, probably for Type XXI boats, was already given up by the OT after the 1944 invasion of Normandy.

As of April 1943, the OT began to strengthen the bunker decks, sometimes to a thickness of 6.1 meters, as the boats kept getting larger. Neither the bunker strengthening nor the catching grid could be finished until August 1944. The catching grid consisted of vertical members some 3.8 meters high, to which ribs about 1.5 meters high were attached at intervals of 1.5 meters. These were supposed to make bombs that struck them explode above the actual upper deck. In the same year, the bunker roof was equipped with three Flak towers for 4 cm Bofors guns, and for the naval guns' fire control, an FuMG 39 T radar unit was installed.

On August 5, 12 and 13, 1944, the 617th Squadron of the RAF attacked the bunker in Brest with Tallboy bombs.

The side entrance to the bunker could be closed with steel doors weighing a ton or more. Over the door was the motto "Through Battle to Victory"...

These bombs were 6.4 meters long, weighed 5.4 tons, and contained 2.3 tons of explosive. In the bomb attacks, the RAF scored nine direct hits on the bunker, five of which tore holes in the bunker roof.

Early in August 1944 the VIII US Corps surrounded the fortress of Brest, capturing it on September 18 after heavy fighting. The last U-boat, U 256, had left Brest on September 4, and arrived at Bergen on October 17. Today the U-boat bunker is still partially used by the French Navy.

The rear of the U-boat bunker, with the thickening and the catching grids visible on the roof. In the structure on the right side, the heating system and administrative offices were housed.

The Brest bunker sustained some serious damage in August 1944. Today a roof has been built over it to keep rainwater out.

Torpedos were brought from underground storage to the U-boat bunker on a narrow-gauge railway.

The U-boat bunker in its present condition, almost unchanged externally. The original doors of the side entrance are still in place. A concrete Flak-gun position can be seen above the entrance to the wet boxes.

The northwest side of the U-boat bunker. The structure included a spare-parts depot.

Below: A drawing of the U-boat bunker in Brest.

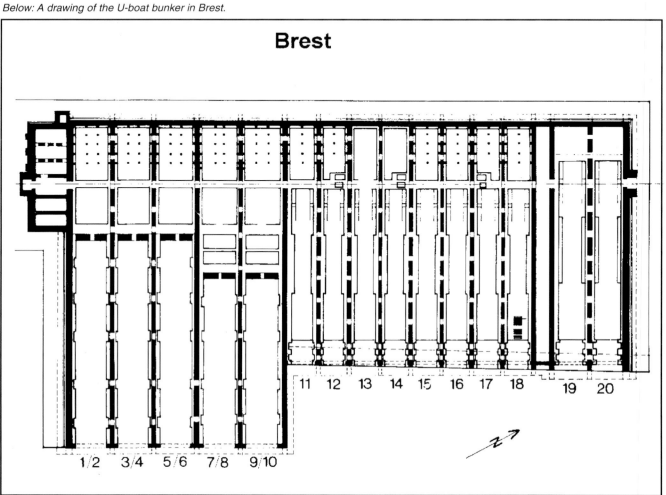

Brest

11 12 13 14 15 16 17 18 19 20

1/2 3/4 5/6 7/8 9/10

Lorient

From 1940 to 1944, Lorient was the main support point of the German U-boats on the Atlantic coast. A few days after the harbor was captured by the Wehrmacht on June 21, 1940, the Navy began to set up the harbor and the arsenal as a U-boat support point, and on July 7 the first U-boat, U 30, under the command of Kptl. Fritz-Julius Lemp, arrived at Lorient.

In the harbor there was a winch for fishing boats, ending in a turntable with 200-ton capacity. Since this capacity was not enough for the U-boats, the facility was enlarged and strengthened by the naval shipyard, so that it was sufficient for Type II U-boats, which weighed 250 tons. On August 19, 1940, U 59 became the first boat to be winched into a dock position and overhauled.

On two of the six dock spaces, the OT built pointed-arch bunkers, so-called cathedral bunkers, with a maximum wall thickness of 1.5 meters. The work on these bunkers, 81 meters long, 16 wide and 25 high, was finished in May 1941, and the west bnuker had a flak gun position installed on its roof. In June 1941 the 2nd U-Flotilla was transferred from Wilhelmshaven to Lorient, and in the following year the 10th U-Flotilla was established there. The B.d.U., Admiral Dönitz, also set up his headquarters in Lorient, in a villa in Kernevel opposite the peninsula of Keroman.

In April 1941 the work on a two-box wet bunker near the naval arsenal had begun; it was 145 meters long, 51 wide and 15 high. The boxes were 99.5 meters long and sufficed to hold types VII and IX boats.

In February 1941, the OT began. on account of the rocky subsoil, to build a dry bunker on the peninsula of Keroman. This facility, consisting of two bunkers, K I and K II, was to have space for twelve U-boats. A winching apparatus was erected, consisting of a very steep flat surface with a wedge-shaped wagon on which a dock wagon was lowered into the water. The boat moved into it and was lifted in the dock wagon, then drawn up the actual slope and moved by an electric locomotive onto a staging beside a bunker box. Every position could be closed off by an armored gate, and docking took about 35 minutes. On the other hand, in a normal drydock where water was pumped out, the process took 4 to 5 hours. The Keroman facility's only drawback was that the large area of the staging was unprotected from air raids, but this was accepted because the facility could be repaired in a short time, thanks to a supply of spare parts. After a building time of only seven months, Keroman I could be put into service in September 1941, with Keroman II following three months later. In 1942 the OT built an extension onto each bunker to provide space for power generators as well as water and fuel tanks. The deck strength of K I and K II was 3.5 meters; because of the war situation in 1944, a catching grid was not built. The expansion of the U-boat war required the building of additional spaces, and so in October 1941 the Keroman III wet bunker was begun. This included seven boxes with berths for 13 boats, plus an additional workshop section.

The Keroman III bunker in Lorient consisted of seven boxes, two for one each, four for two each, and one for three U-boats. On the roof, three Flak towers for 4 cm Bofors guns were built of concrete. At the left side of the picture, the winch to the Keroman II bunker can be seen.

The K III facility was 170 meters wide. 138 long and 20 high. In January 1943 the bunker could be put into service, though the two drydocks could be used only as wet boxes, since a watertight gate could not be built and a calculating figure had been wrong. Every box had two desktop cranes with five-ton load limits, while two boxes even had a 30-ton crane.

As of July 1943, the OT began to build a second deck of 2-meter concrete to protect K III. The workshop area, on the other hand, received only a catching grid for its protection.

A planned expansion of the bunker facilities, designated K IV, to hold 24 Type XXI U-boats, could not be carried out because of the war situation. Opposite Keroman II the OT built six torpedo bunkers, which were given the disguised names of "Jaguar, Iltis (Polecat), Leopard, Luchs (Lynx), Tiger and Wolf". From these bunkers, 40 meters long and 23 meters wide, torpedos could be brought over a narrow-gauge rail line directly into the U-boat bunker. From the beginning of 1942 to July 1943, up to thirty U-boats were based at Lorient; the fact that the crews, after week-long missions in very cramped quarters, naturally often "let off steam" was accepted by the Wehrmacht units stationed on land.

On August 6, 1944, 28 Lancasters attacked the Keroman bunkers, dropping eleven Tallboy bombs, of which only one scored a direct hit on Keroman III. The bunker deck was 7.5 meters thick at that point, and except for an outward bending of the sheet steel on the underside of the deck, there was no real damage done. One day later, the first American armored troops reached the fortifications of Lorient, and soon the city was surrounded. On September 5, U 155 became the last U-boat to leave the bunker. U 123 and U 129, which could not be repaired, were left behind. U 123 was put back into service by the French Navy in 1947

as S-10 Blaison. Today the U-boat bunkers are still partially used by the French Navy, and the headquarters of the B.d.U. in Kerneval is now the residence of the commanding admiral.of the naval harbor of Lorient.

U 106 lies at the quay of the naval arsenal in Lorient in September 1941.

One of the two cathedral bunkers by the fishing-boat winch; the windows in the armor-plated doors were cut only after the way. On Jamuary 29, 1943 the 2 cm Flak gun position was nearly hit during a bombing raid. Obergefreite Uher dragged his two badly injured comrades down over the arched bunker wall during the attack, so that they could be taken to the sickbay.

A boat of the 10th U Flotilla returns to Lorient after a mission in the summer of 1942.

The Keroman III wet bunker.

Keroman II, seen shortly before it was finished. In front is the moving wagon on the staging. The U-boat boxes can be closed off by armored doors. The outer box on the left side served as a protected garage for the moving wagon; over it were barracks for 1000 men.

Above and below: K II in its present condition, used regularly by the French Navy for maintenance work.

14

U 30 leaves the harbor of Lorient at the end of June 1940, after being overhauled, and sets out on a mission in the Atlantic.

The main entrance to Keroman I.

After the fortress of Lorient was surrounded, the U-boat bunkers were used as sheltered barracks and workshops after U 155 became the last boat to leave Lorient on September 5, 1944. In these workshops, armored trucks and light tanks were built to defend the fortifications.

German soldiers march past the Keroman II bunker to go on duty at the fortifications.

Left: The entrance of the "Jaguar" torpedo bunker.

Left: A motto on the wall of the "Jaguar" bunker, which is used today as a storehouse for a paint firm and can be visited. ("Dare the boldest, strive for the highest, bear the heaviest: a German life.")

Right: The moving stage between the dry bunkers was never seriously damaged despite heavy bombing, and remained in service until the war ended.

At the edge of Lorient, the bow of a French submarine stands today as a monument to the long seafaring tradition of the city. In the background, the "Luchs" (Lynx) torpedo bunker can be seen.

Torpedos are stacked in noe of the supply bunkers.

St. Nazaire

In St. Nazaire the 7th U-Flotilla, named the Wegener Flotilla after Admiral Wolfgang Wegener, was based as of Junq 1941. In February 1942, the 6th U-Flotilla followed, the Hundius Flotilla, named after the World War I U-boat commander and Pour le Me/*rite bearer Paul Hundius. U 48, the most successful U-boat of World War II, belonged to the 7th Flotilla. On thirteen missions, the boat, with Commanders Schultze, Rösing and Bleichrodt, sank a total of 310,407 BRT.

In March 1941, the OT began to build a U-boat bunker in the south harbor of St. Nazaire. After just four months, the first three boxes could be used. By June 1942, all fourteen boxes, eight drydocks and six wet boxes could be put into use. For the bunker, the OT used some 480,000 cubic meters of concrete; it was 295 meters wide, 130 meters long and 18 high.

In 1943 a concrete extension, 121 meters long and 22 high, was built onto the eastern part of the bunker to provide space for bunkers.

In June 1943, the OT began to construct a second deck, at first 3.85 meters thick. When the work was stopped in August 1943, about 90% of the deck and 30% of the catching grid had been finished.

Opposite the box entrances, a bunkered lock entrance was built, starting in August 1942. This structure, 155 meters long and 25 wide, was equipped with numerous short-range defenses, as was the U-boat bunker itself. On the roof was a six-loophole turret and 4 cm Bofors Flak gun positions. After the surrounding of St. Nazaire in August 1944, the Navy disbanded the 6th U-Flotilla and transferred the 7th Flotilla to Norwegian harbors. In the coming months, several U-boats supplied the fortifications with medical supplies, ammunition and mail. On April 24, 1945 U 510, coming from Batavia, became the last boat to enter the bunker; after the capitulation it was taken over by the French Navy and put into service in 1946 as S-11 "Bouan". Today the bunker is completely in private use and can be seen very easily. At the entrance it still says, "Show identification unasked", and many German inscriptions can be read inside the bunker as well. The wall inscriptions, such as "U-Hein, U-Seepferd und U-Glückauf", with the signatures of several crew members, on the box walls have a strange effect on the visitor today.

Left: U 455 of the 7th U-Flotilla returns to the St. Nazaire support point after a mission in the autumn of 1942.

Below: A VII C boat lies before the St. Nazaire bunker in the summer of 1943. The work on the second deck had already been begun at this point in time.

The building site of the U-boat bunker, seen in the spring of 1941. On June 30, after four months of construction, the first three boxes could be used. U 203, under Kptl. Mützelburg, was the first boat to enter the bunker.
Below: Three overhauled VII C U-boats leave the bunker in the spring of 1943.

An IX C U-boat in Box 13 of the St. Nazaire bunker.
The same box fifty years later; the U-boat is gone, but practically nothing else has changed.

The harbor of St. Nazaire, seen from the air. The Normandy Lock, the main target of British commando operations in 1942, can be seen at the right. In the middle is the lock bunker, with the U-boat bunker to the left of it.

Right: U 85 enters the south harbor at St. Nazaire after a mission. The boat was lost near Cape Hatteras on April 14, 1942.

Two boats of the 7th U-Flotilla in the lock at St. Nazaire.

The landward side of the bunker, seen in July 1942. The balloons are tethered in front of the box entrances as a defense against low-flying planes.

Right: The bunker in its present condition; its camouflage paint can still be seen on the shaded walls.

Construction work on the bunker deck in 1970.

For security reasons, the U-boat boxes of the bunker could only be entered by crew members and selected shipyard workers, who were given numbered plates for admission.

Right and below: Lettering on the box walls.

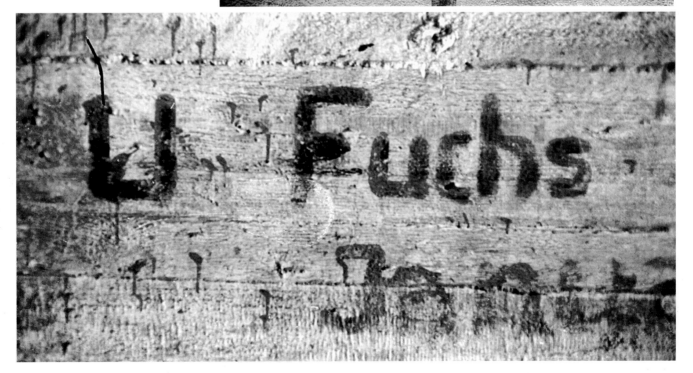

This workshop and supply bunker at the St. Nazaire naval shipyard was torn down after the war.

Dock 6 of the St. Nazaire bunker in its present condition. The closing and lifting gates were removed after the war. On the quay, what remains of the four pumps that served to empty and fill the docks can be seen.

U 510 was in Dock 4 of the St. Nazaire bunker in May 1945 was was being readied to have a snorkel installed. After the surrender, the French Navy took over the boat.

The U-boat bunker fifty years later. In the summer of 1943, the roof was given a second deck and a Flak gun position.

La Pallice

In October 1941, La Pallice became the home port of the 3rd U-Flotilla, the "Lohs" Flotilla, named after the successful U-boat commander and Pour le Mërite bearer of World War I, Oberleutnant zur See Lohs.

The work on a U-boat bunker had begun in April 1941, and in October the first two boxes could be completed. The five drydocks followed in November. In April 1942, the OT began to extend the bunkers by another three docks. To do so, some 200 meters of soil had to be removed from in front of the bunker in order to flood the boxes. After the work was finished, the bunker had three wet boxes and seven drydocks. Boxes 1 and 2 had a length of 92.5 meters and a width of 17; Docks 3 to 7 were 92.5 meters long and 11 wide; Docks 8 to 10 were 100 meters long, and the last was 16 meters wide.

The structure was now 195 meters long, 165 meters wide, and 19 meters high. The deck thickness varied, with 7.3 meters over boxes 1 to 5, and 6.5 meters over boxes 6 to 10. The work on the decks and the catching grid was still begun, but in August 1944, shortly before the encircling of the fortress of La Rochelle-La Pallice by American troops, the OT called a halt to it.

At the northwest of the bunker, the OT constructed a concrete addition as a powerplant. This tower, 30 meters wide, 50 long and 25 high, housed the electric power generators and controls. On August 9 and 18, 1944, the bunker of La Pallice was the target of Tallboy bombs dropped by the 9th and 617th RAF Squadrons. Of the eighteen bombs dropped, four scored direct hits on the bunker, causing two breakthroughs. On September 10, 1944, U 382 became the last boat to leave La Pallice; U 766 was still in the bunker, as it could no longer be made ready to depart, and the 3rd U-Flotilla was disbanded. The commander of the fortress, Vice-Admiral Ernst Schirlitz, made an agreement with the besieging troops that, if there was no attack on the fortress, no destruction would take place in the harbor. Thus the harbor, vital to the city, remained unharmed.

A returning boat of the 3rd U-Flotilla is led into La Pallice harbor by minesweepers.

The U-boat bunker in the harbor of La Pallice, seen from the lock bunker.

The first boxes of the bunker were finished in the spring of 1942.

The bunker of La Pallice as it looks today. In the foreground are the three added boxes. This part is in civilian use today and is open to visitors.

The camouflage paint on the box entrances can still be seen clearly today. The armored gates of boxes 1 to 9, installed in the autumn of 1943, were used as scrap metal after the war.

The north side entrance of the La Pallice bunker; the overhanging roof was to prevent bombs from hitting the bunker wall.

La Pallice

1-10 Boxen

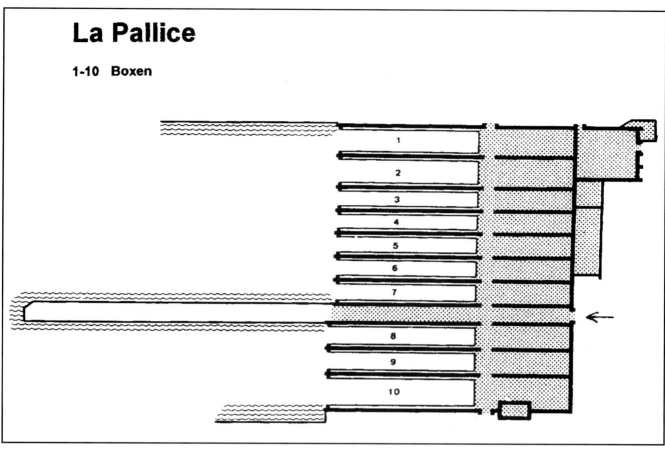

The layout of the bunker at La Pallice.

Between the first and second sections, there was an intermediate box with a 200-meter pier. In the box, the OT laid a double rail line. On the ceiling is sheet metal, which prevented concrete from falling down in case the deck was damaged by a bomb.

The entrance to the power house of the bunker, in which the power generator was housed. All the entrances could be closed by heavy steel doors.

The entrance to the La Pallice lock bunker. In June 1942, the OT began the work on the bunker, which was 167 meters long, 26 wide and 14 high; it was put into service in March 1944. But the bunker lock was used only as of August 1944, after the destruction of the main lock in a bombing raid. The first boat to pass through the bunker was U 963.

A look at the south side of the lock bunker. The inscription "Wir bauen für den Sieg" (We build for victory) was put there for the filming of the film "Das Boot".

In the vicinity of the U-boat bunker the OT built a row of smaller structures, including this torpedo bunker.

The OT built three more torpedo bunkers east of the U-boat bunker in 1943. There was a rail link for transport between these supply bunkers and the U-boat bunker.

Bordeaux

As of September 1940, Bordeaux was the home port of the Italian Betasom U-Flotilla. The boats used Basin I and two drydocks. In the summer of 1941, the German Navy also decided to station U-boats in Bordeaux. For this, the OT constructed a U-boat bunker in Basin 2, beginning in September of that year. The bunker was 245 meters long, 162 meters wide and 19 meters high. Since two boats of the larger types, such as XIV, X B and IX D, were to be housed in the wet boxes side by side, and the U-tanker of Type XIV had a stern width of 9.35 meters, the OT set the width at 20 meters. On January 17, 1943, U 178 of the newly established 12th U-Flotilla entered the bunker, and in May the construction work was finished for the time being.

As of mid-1943, the OT began to thicken the decks of the bunker roof from 3.5 to 5.6 meters, and the greater part of the bunker was also fitted with a catching grid. In all, more than 600,000 square meters of reinforced steel were used in the U-boat bunker.

In the vicinity of the structure, the OT erected an oil-fuel bunker, measuring 80 x 38 x 15 meters. At the end of August 1944, as American troops approached Bordeaux, the Wehrmacht gave orders to evacuate the harbor. On August 25 the last two U-boats, U 535 and U 857, left Bordeaux, and in October the 12th U-Flotilla was disbanded officially.

The bunker and the harbor facilities were supposed to be blown up on command, through the considerate actions of Uffz. Heinz Stahlschmidt, who was in charge of the explosives and the demolition, this was avoided. Stahlschmidt made contact with the local Resistance and turned the undamaged harbor over to them on August 28, 1944. Today the bunker is in civilian use and in very good condition.

The Italian U-boat "Barbarigo" passes through the outer lock from the Gironde River into Basin I. On the pier, Admiral Angelo Parona greets the returning boat.

The front of the Bordeaux U-boat bunker, on which the box numbering can still be seen today.

A look at Boxes 8 to 11 in their present condition, with the catching grid visible on the roof of the bunker.

Left: The "Barbarigo" in Basin 1, with two Marconi-class boats at the pier in the background.

In the boxes there were large ventilators which removed the Diesel exhaust of the boat motors during test runs.

Here foreign guests are visiting the bunker, which has been covered with camouflage nets.

Drydocks 9 to 11 had a usable surface of 90 meters, while Docks 5 to 8 measured 96 meters and the four wet boxes, 1 to 4, were 102 meters long.

The entrance to the Bordeaux bunker, with the steel doors still present, fifty years after the war's end.

Bergen

As of the autumn of 1940, German U-boats entered the Norwegian harbors, particularly Bergen and Trondheim, to take on fuel oil and have minor repairs made. In the spring of 1941, the Navy began to make plans for a U-boat bunker in Bergen; a small bay on the west of the city was chosen as the location. The bunker, with a size of 130.5 x 143 meters, was to have a capacity of nine boats, divided among three wet and three dry boxes. The seventh box was intended as a storeroom for fuel, as well as for oxygen production machinery and generators. Over the boxes, a mezzanine with a deck height of 3.7 to 3.9 meters was to be built, in order to gain additional space for storerooms and a cafeteria. Along with the extra space, this meant an increase in safety against bombs for the U-boat boxes. The included floor of the mezzanine, 1.35 meters thick, was capable of supporting considerable parts of the main deck, in case the latter should be penetrated by bombs. To be sure, a mezzanine was only to be extended over boxes 1, 2 and 7.

The construction of the bunker was not overseen by the OT, but by the Harbor Construction Agency of the Bergen Naval Shipyard. which in turn gave contracts to various construction firms. Soon after the work began in November 1941, the first difficulties surfaced. Because of the low population density of Norway, only a few workers were available, and at first the Navy settled penal companies in isolated places on the building site. Another problem was the poor road and rail connections in Norway. At times very laborious preparations had to be made, since the building materials had to be brought over long distances. In addition, the climatic conditions in Norway influenced the construction greatly.

In its final form, the deck of the bunker was to have a thickness of six meters of reinforced concrete plus a granite covering one meter thick. By the end of the war, though, only Box 2 had a reinforced concrete deck of the planned thickness. The other boxes had decks of only 3.5 to 4 meters. The main and mezzanine decks were built with the help of tensile concrete supports; the supports over the dry boxes measured 15 meters and weighed 11 tons. The supports over the wet boxes were 21 meters long and weighed 21 tons.

The openings of the dry boxes could be closed by means of four-part iron doors 33 mm thick, while the wet boxes remained open. Every box had a three-ton mobile crane; Box 3 even had a 30-ton crane to remove Diesel engines. In mid-1944 the first U-boats entered the bunker, and as of that December the whole complex was functioning.

The city of Bergen and the U-boat bunker were targeted by the RAF on October 4 and 29, 1944, as well as on January 12, 1945. In every attack, the bombers achieved direct hits on the deck, which was still under construction, and the building-site facilities. This caused delays in the further construction of the bunker. In the attack on January 12, 1945, the British bombers dropped 24 Tallboys, two of which penetrated the deck. Major damage to the boxes, though, was prevented by the completed mezzanine. After the war, the British blew up part of the U-boat bunker. In 1949 the Norwegian Navy put the drydocks back into operation, filled the destroyed Boxes 4 to 6, and extended them to form a quay. The workshops in the bunker are also still in use today.

Three U-boats of Type VII C of the 11th U-Flotilla lie in the bay at Bergen, ready to depart. Repairs and overhauling had to be carried out completely in the open air until mid-1944. After that, part of the naval shipyard could move into the "Bruno" bunker. All the same, work still had to be done outdoors because of overcrowding in the support point and shipyard. As a result, two boats, U 228 and U 993, which were in a floating dock, sustained severe damage in an RAF attack on October 4, 1944, and had to be scrapped.

The construction of the concrete sidewalls of the "Bruno" bunker was finished in the summer of 1943.

The front of the "Bruno" U-boat bunker. The three drydocks can be seen at the left. The blown-up parts were filled in by the Norwegian Navy and used as a quay.

The "Bruno" U-boat bunker in its present condition.

A VII C boat of the 11th U-Flotilla returns from a mission. The five pennants on the masthead show how many tons of shipping were sunk.

The boxes of the "Dora I" bunker built up after the war; today the bunker is used for civilian storage and parking.

Trondheim

In the shipyards of Trondheim, the Navy repaired the boats of thr 13th U-Flotilla, and as of 1941 more and more of the U-boats from the northern support points of Narvik, Kirkenes and Hammerfest.

In the same year, the OT began to build the first of two U-boat bunkers in the harbor area of Trondheim. This one, with the code name of "Dora I", measured 153 x 105 meters and consisted of five boxes. Three of them were planned as drydocks for one boat each, and two as wet boxes for two boats each. In the bunker there were also workshops, and a railroad spur was also built.

The deck thickness was 3.5 meters of reinforced concrete, the outer walls were three meters thick and the inner walls 1.25 to 2 meters. On July 20, 1943, after a building time of 27 months, the bunker was opened. The OT began to build the second bunker, "Dora II", in January 1942. It measured 168 x 102 meters and was to have two dry and two wet boxes. When the war ended in May 1945, only 65% of the work had been finished. After the war, the British blew up part of the "Dora II" bunker, while "Dora I" was at first used as a submarine bunker by the Norwegian Navy. From 1955 on, the Navy turned the bunker over to civilian use as a storehouse. The city of Trondheim built a parking garage on the roof in 1988.

The partly completed "Dora II" bunker in the harbor of Trondheim. The deck design with Melan supports can be seen here. These prefabricated iron supports were up to 29 meters long, depending on the width of the box. They were placed at intervals of 1.5 meters, with round iron bars between them, and could also be covered with concrete.

What remains of "Dora II" is used today as the work- and storeroom of a shipyard, and is open to visitors.

The side entrance of the "Dora II" bunker, with short-range defense positions.

The "Valentin" U-Boat Factory

In the spring of 1943, plans were begun for a production bunker, code-named "Valentin", in Farge near Bremen. It was planned that the electric Type XXI U-boat would be built there. With this first real underwater ship, the U-boat war was to be given another turn in Germany's favor. By means of a snorkel, the boat would be supplied with fresh air without surfacing, and its exhaust gases could be discharged in the same way. This boat could be built quickly and in large series, and production began at various shipyards as of June 8, 1943. For the first time, the boats were not, as before, completely built at a shipyard, but prefabricated in eight sections at various production facilities and only assembled at the end. This resulted in a great saving in work and time. The boat required only 260 hours to be finished, instead of 460 hours as before. In the shipyard where it was assembled, the boat required only eight instead of forty weeks.

Because of the strategic bombing war of the Allies on Germany, in which shipyards, factories and railroad lines were often destroyed, with 173 bombing raids flown against Bremen alone, there had been long delays in the manufacture of U-boats.

The eight individual sections of the XXI boat were to be made on an assembly line in the "Valentin" bunker. In stages 1 to 3, the individual sections, including the tower, were to be welded; in stages 4 to 8, the interiors were to be added, in stages 9 to 11 the periscopes installed, in stages 12 and 13 the final check and launching. The boat would then run along a 10-meter-long canal into the Weser. The bunker was also to have a powerplant to produce energy, workshops and supply storage space.

With the beginning of Type XXI production at the shipyards, the final measurements for the "Valentin" bunker could be established. The work was assigned by the Weser Naval Construction Office to various firms. More than 12,000 men worked at the building site, of whom 4000 were foreigners of various nationalities, 500 Russian prisoners of war, and 2000 prisoners from the Neuengamme concentration camp. The rest of them were naval construction workers, civilian employees of the construction firms, and guards from the SS and the Navy. In May 1944, the OT took over naval construction, which once again intensified the work on the "Valentin" bunker. Even so, the bunker could not be finished, as planned, in the autumn of 1944. On account of the bad food and inhuman treatment, some 4000 of the prisoners amd POWs died before the work was halted on April 7, 1945. At this point the bunker was about 90% finished.

In two air raids on the "Valentin", on March 27 and 30, 1945, the Allies used special bombs to break through the deck three times. This caused some damage, of course, but the main effect of the bombs was wasted in the gigantic interior of the bunker.

The "Valentin" bunker, with its 49,276 square meters, was the second largest U-boat bunker of World War II. It used 220,000 tons of cement, 27,000 tons of steel, and one million tons of additional material. Its cost was 120 million Reichsmark.

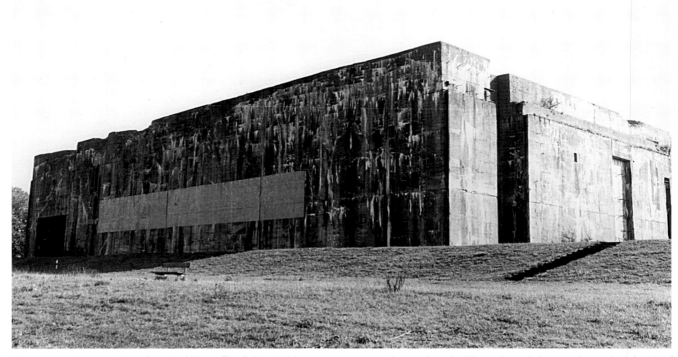

The "Valentin" bunker, seen from the Weser. The finished U-boats were supposed to run into the Weser through the opening at the left side of the picture.

Series production of pressure units at a German shipyard in 1944.

Through this side entrance to the bunker, the individual sections of the XXI U-boats were to be moved into the bunker. An extension for storage of delivered sections was also planned for this place.

The "Valentin" bunker as seen from the landward side. In the foreground is the monument, erected in 1985, to the many concentration-camp and war prisoners who had to work at the site, generally under inhuman conditions.

This cutaway drawing of the "Valentin" bunker shows the sequence of U-boat production that was planned for 1945.

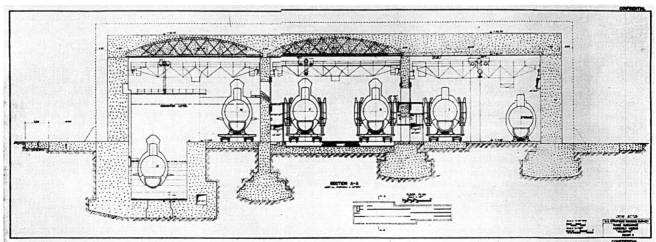

Two VII C U-boats lie in the St. Nazaire bunker, ready to set out.

Right: The bunker at Brest fell into the hands of the Allies mostly undamaged and was immediately put to use by the French Navy.

ASSIGNMENT OF U-FLOTILLAS TO THE MENTIONED BUNKER SUPPORT POINTS

1914-18

Flanders	1st U-Flotilla	St. Nazaire*	6th U-Flotilla
	2nd U-Flotilla		7th U-Flotilla

1940-45

Brest*	1st U-Flotilla	La Pallice*	3rd U-Flotilla
	9th U-Flotilla	Bordeaux*	12th U-Flotilla
Lorient*	2nd U-Flotilla	Bergen	11th U-Flotilla
	10th U-Flotilla	Trondheim	13th U-Flotilla

* After the Allied conquest of France after June 1944 and the loss or closing of the Atlantic support points, the U-boats were transferred from these harbors to Norway or Germany.

Overview of the U-Boat Bunkers in World War II

Place	Code Name	Dimensions in Meters	Deck thickness	Number of Boxes	Boat Capacity	Building Time
Bergen	Bruno	130.5 x 143	6 m	6	9	1941-44
Trondheim	Dora I	153 x 105	3.5 m	5	7	1941-43
Trondheim	Dora II	168 x 102	3.5 m	4	6	1942-45
Kiel	Kilian	176 x 79	4.8 m	2	12	1941-43
Kiel	Konrad	163 x 35	3.5 m	1		1943-44
Hamburg	Elbe II	137 x 62	3 m	2	6	1940-41
Hamburg	Fink II	151 x 153	3.6 m	5	15	1940-44
Helgoland	Nordsee III	156 x 88	3 m	3	9	1940-41
Bremen	Valentin	426 x 97	7 m	-	-	1943-45
Brest	-	192 x 333	6.2 m	15	20	1941-42
Lorient	Dombunker	81 x 16	1.5 m	1	1	1941
Lorient	Scorff Bunker	145 x 51	3.5 m	2	4	1941
Lorient	Keroman I	403 x 146	3.5 m	5	5	1941
Lorient	Keroman II	403 x 146	3.5 m	7	7	1941
Lorient	Keroman III	170 x 138	7.5 m	7	13	1941-43
St. Nazaire	-	295 x 130	7 m	14	20	1941-42
La Pallice	-	195 x 165	7.3 m	10	13	1941-43
Bordeaux	Orion II?	245 x 162	5.6 m	11	15	1941-43

The U-boat Bunker of St. Nazaire in 1980.